Christian Röse

Subsidies in the European Agriculture - Using the Example of the Südzucker AG

GRIN Verlag

Bibliografische Information der Deutschen Nationalbibliothek:

Die Deutsche Bibliothek verzeichnet diese Publikation in der Deutschen National-
bibliografie; detaillierte bibliografische Daten sind im Internet über http://dnb.d-
nb.de/ abrufbar.

Impressum:

Copyright © 2010 GRIN Verlag GmbH
Druck und Bindung: Books on Demand GmbH, Norderstedt Germany
ISBN: 978-3-640-89133-7

Dieses Buch bei GRIN:

http://www.grin.com/de/e-book/170349/subsidies-in-the-european-agriculture-
using-the-example-of-the-suedzucker

GRIN - Your knowledge has value

Der GRIN Verlag publiziert seit 1998 wissenschaftliche Arbeiten von Studenten, Hochschullehrern und anderen Akademikern als eBook und gedrucktes Buch. Die Verlagswebsite www.grin.com ist die ideale Plattform zur Veröffentlichung von Hausarbeiten, Abschlussarbeiten, wissenschaftlichen Aufsätzen, Dissertationen und Fachbüchern.

Besuchen Sie uns im Internet:

http://www.grin.com/

http://www.facebook.com/grincom

http://www.twitter.com/grin_com

EUB

SUBSIDIES IN THE EUROPEAN AGRICULTURE
— USING THE EXAMPLE OF THE SÜDZUCKER AG

Hanze University Groningen
APPLIED SCIENCES
International Business School

Individual Project Report

Christian Mirko Röse

IBS Groningen
Erasmus Exchange Program

2009-12-26

Table Of Contents

1. Introduction

In between the worldwide system of agricultural activities, the European Union plays a very important role. As they have both, a huge consumer market and as well as an enormous production capacity, the whole world has a look on what kind of politics they make and how they act. A big change in the subvention funds could have tremendous impacts on the world economy.

The European Union spends an enormous amount of money per year to subsidize its agriculture in order to keep the production of their agricultural products within their own acreages and of course to safeguard employment within the European agricultural branch.

As we see Europe as one country, it would have the fourth largest consumption of sugar and the third biggest sugar production. Germany in this case is number two in Europe in terms of the sugar production and the *Südzucker AG* is therefore the biggest in Germany. Because the amount of the payments is primarily tied to the size of the enterprise, consequentially, European the German grocery-producer is also the main beneficiary of the European agriculture subsidies funds.

In the following, first of all the enterprise *Südzucker* will be described briefly. After that, an explanation of the European agricultural policy, including its attitudes concerning subsidies will be given, before it comes to an overview of the EU sugar policy. After all, I will broach the issue of how *Südzucker* should act in the future in order to save its position considering these overall policies.

2. The Enterprise: *Südzucker AG*

The *Südzucker AG*, based in Mannheim, is a global operating German food group with 18,000 employees. Its main segment is sugar. Their turnover per year is EUR 5.9 billion and with their yearly sugar production of 4.2 million tones, they are leader on the European sugar market and one of the largest food companies in Germany. As a stock corporation, *Südzucker* is of course member of the German MDAX. The dividend was paid out 76 million Euros (0.40 Euros per share). The cash flow amounted to 503.8 million Euros. Because a third surprisingly strong quarter in fiscal year 2008/09 amounts to operating earnings in the first three quarters to 184 million euros, compared to 176 million Euros a year earlier. Group sales in the first three quarters increased by 5% to 4.6 billion Euros. (FAZ, 2009) At the same time the *Südzucker AG* was, in 2008, with more than 34 million Euros, the biggest German recipients of EU farm subsidies. (Handelsblatt, 2008) These agricultural subsidies are from the European Agricultural Guarantee Fund, financed by taxes, on production from sugar beet farmers and the sugar industry. (Südzucker , 2010)

The *Südzucker AG* goes back to the "South German sugar-AG", which emerged from a 1926 regional concentration of sugar factories (including sugar factory AG Frankenthal, Baden Society for sugar manufacture, Mannheim, sugar factory Stuttgart AG, Stuttgart-Bad Cannstatt. The official predecessor, the Süddeutsche sugar-AG is the sugar factory Frankenthal AG. The company acquired since 1996, many sugar factories in Eastern Europe, particularly in Poland, and then rose to become by far the largest sugar producers in Europe. In 2005 5.2 million tons of sugar were produced (corresponding to a share of the EU-25 sugar production of 21.8%). On 25 June 2009, the board of *Südzucker AG*, with the approval of the supervisory board, determined an unsecured convertible bond. The bonds will be issued by Südzucker International Finance BV, a 100% Dutch subsidiary of Südzucker, guaranteed them and is convertible into existing or new shares of them as well. The bonds are offered only to institutional investors outside the U.S. for sale. The proceeds from the sale of convertible bonds will be used for general corporate purposes. (Finanznachrichten, 2009)

As spokesman for the board acts Wolfgang Heer from Ludwigshafen, chairman of the board is Hans-Joerg Gebhard from Eppingen, who is also chairman of the Association of South

German sugar beet farmers eV. The remuneration of the Managing board and the Supervisory board together amounted in 2008-09 to 6.4 million Euros. (Südzucker, 2010)

Südzucker AG is only in three countries of Europe, factories in Germany, Poland and Moldova, established under this name. In Poland and Moldova was established in each subsidiary Südzucker Polska SA and Südzucker Moldova SA.

The company is 55% owned by the Southern German sugar beets exploitation association in Stuttgart, representing in the 30,000 sugar beet farmers in the region. The ZSG Netherlands BV, Amsterdam, Netherlands, with 10% and the Leipnik-Lundenburger Invest Beteiligungs AG, Vienna, owns 2.3%. The rest of the shares are in free float.

3. The general current situation of EU agriculture policy

To understand the consequences for an end-producer like Südzucker, an outlook of the general current situation of the common agricultural policy should be given, also in order to give a clear view on the basic conditions for the producers in the first production step.

The share of agricultural spending in the EU budget is declining, but with 43% (about 56 billion Euros) it is still the largest single budget. In the year 1977, the proportion was 76%. The largest part of expenditure causes the CMOs and their related agricultural subsidies. The EU guaranteed the producers of agricultural products minimum prices. Since these were lowered several times in the past, they receive direct compensation since Agenda 2000, largely independent of the produced quantity. (European Commission, 2009); (Auswärtiges Amt Deutschland (Federal foreign office), 2009)

The objectives of the Common Agricultural Policy were set out in Article 39 of the Treaty of Rome: (European Parliament, 2000)

1. To increase agricultural productivity by promoting technical progress, rationalization of agricultural production and the optimum use of production factors, especially labor

2. The agricultural community, in particular by increasing the per capita income of the agricultural working people, ensure an adequate standard of living

3. To stabilize markets

4. To ensure the supply

5. To ensure that supplies reach consumers at reasonable prices

In Article 34 the creation of a common organization of agricultural markets (CMO), shall be determined, which has, depending on the products, featured on of the following organization: (hri.org)

1. Common rules on competition

2. Compulsory coordination of the various national organizations

3. A European regime

In 1962, the common organization of agricultural markets defined three principles for the common agricultural market. (European Navigator ENA, 2006)

1. Unity of the market: This means the free circulation of agricultural products within the member states, should be applied throughout the EU, the same instruments and mechanisms for the organization of the internal market

2. Community preference: This means that the EU agricultural products have a priority and a price advantage over imported products, which means also the protection of the internal market economy products from third countries and from major fluctuations in the global market

3. Financial solidarity: All expenditure under the CAP will be borne by the community budget.

The Common Agricultural Policy recognizes the structure of peasant agriculture and the structural and natural disparities between the regions and aims at a gradual adjustment of the conditions.

Trade within the EU (intervention prices): Every year, the EU sets a minimum price/intervention prices for certain agricultural commodities. If the market price falls below the minimum price/intervention prices, the EU buys from producers of these products. This is called a support buying. Through this support buying, on the one hand, producer prices are stabilized and taken other surpluses from the market. This will prevent further subsidence of the market price. The purchased products will be centrally stored and, depending on the market, then sold again. This price and purchase guarantee, however, encouraging overproduction. Through the CAP reforms of 1992 and 2003 and the Agenda 2000 intervention prices have been drastically reduced and replaced by income support.

The production costs for almost all agricultural products that are produced in the European Community are far above the level of world prices. Therefore they are not competitive. To prevent flooding the European market with imports from other countries, the EU has adopted the threshold price. A vendor from a non-EU country must pay out the difference between the world price and the trigger price to the EU as a sort of duty. This arrangement is called a levy. For non-EU countries it is hard to access the European market. Above all, developing countries are affected by the scheme. Also, consumers in the EU countries are affected because they have to pay higher prices than the world market.

To be competitive on the international market, producers are able to let pay out the difference between the world price and the trigger price by the EU. This means that a farmer can sell his products at the low world market price and still make profit. (Schwarz & Pfeifer, 2006)

On the 26 June 2003, in Luxembourg, the EU agriculture ministers agreed on reforming the Common Agricultural Policy (so-called Luxembourg decisions). The EU Member States implemented these directives into national regulations from 2005. The agrarian reform comes at a time in which developing countries, with efforts in the context of WTO negotiations, gain access to the EU's internal market. (European Commission - Agriculture and Rural Development, 2003)

Targets of the 2003 reform are certainly on the one hand to limit the increase in spendings on agricultural policies after the EU enlargement of 2004. Currently, more than 40% of all EU spending on the Common Agricultural Policy will be applied. On the other hand the measurement to safeguard the environment and consumer protection will be strengthened. Nationally implemented though were only "Decoupling of direct payments", "Cross Compliance" and the "Modulation" of the EU agriculture policy into two pillars.

Decoupling of direct production (European Commission - Agriculture and Rural Development, 2003)

Applied to 2004, the complex system of production-linked direct payments will be converted gradually until 2013 to farm-decoupled direct payments ("farm payments"). This should lead to regionally uniform hectare premiums and greater market orientation. The choice of which product produced by a farmer in the future, will no longer be determined primarily by the level of product-related payments, but will depend on market conditions.

According to the German Federal Ministry of food, agriculture and consumer protection, thus, extensive and ecologically working farms would be strengthen, existing imbalances in the recent promotion, such as the discrimination of grassland sites, would be fixed. Finally, a more transparent and more accepted by consumers system of direct payments would be established and administrative expenses would be reduced.

In Germany, the vertices of the decoupling model in which they are regulated as Article 1 of the Act to implement the reform of the Common Agricultural Policy Implementation Act contained single. The decoupling begins on 1 January 2005. (Agrarrecht)

The process of decoupling includes 3 steps:

Assignment

The decoupled premiums are allocated to farms in the form of so-called entitlements per hectare. For this re-allocation of decoupled premiums to the companies, the basic Regulation provides two options. The enterprises can be assigned individual payment-entitlements, which are: Its historical premium volume in the reference period 2000 to 2002 divided by the cultivated area in this period. This so-called operational model contrasts with the regional model, which provides for identification of regionally uniformed entitlements per hectare. Germany has opted for a combination of both models.

The number of entitlements, the farmer receives a given request is determined basically by the number of hectares of eligible land in 2005.

The value of entitlements is composed of a unit area amount and an individual amount together. The individual amount is derived from certain animal-related bonuses and aid received by the individual farmer in the reference period from the milk premium, payable to the farmer because of its reference quantity of milk of the year 2005/2006, covering parts of the recent tobacco premium) and amounts to the farmer in connection with the reform of the sugar regime and the Common Market Organization for Fruit and Vegetables. The area-sum results in simple terms from the previous regional aid levels for certain crops and a part of the animal premium of the reference period. In Germany, this regional premium volume was applied to eligible land of the year 2005.

Entitlements were generally assigned for all agricultural land with submission of applications for 2005 except of permanent crops, as well for those who have not yet received any gifts, like fruit-producers, vegetables and potatoes-producers as well as areas of horse owners.

The absolute amount of direct payments including the budget for the modulation (see below) is not changed, the extensions will be borne by the recent award recipients. In 2008, the request of entitlements for fruit orchards (permanent crops) as well as nurseries and nursery areas were assigned.

The entitlements are provided with a unique label and recorded in a database. Any entitlement from initial allocation is freely tradable. The exception would apply to payment entitlements allocated from legitimate reasons, or increased by at least 20% in value. The entitlements are activated each year with the exception of set-aside entitlements with each eligible area. Up to and including 2007, decommissioning entitlements could only be activated with surfaces, which were closed down.

Approximation

The above entitlements were remained unchanged until 2009. 2008, a so-called "health check" of the agrarian reform took place, the effects of the reform before were reviewed.

In the phase of aligning, the different entitlements will receive a single payment to claim value until year 2013. Therefore, entitlements that were above the target in the year 2009, are progressively decreased in value, debts that are below target, gradually increased.

From 2013, all entitlements due to the alignment of a region have a uniform amount.

Cross Compliance

To receive funding, the farmers must meet certain basic requirements for the production. This includes environmental and animal welfare and food covers and feed safety. The novelty is, that in cases where this reduction, which is a already existing EU standard, the direct payments are reduced (for first-time violations not more than 5%) or deducted in full for willful violations, in extreme cases. In addition, regulations on soil protection and the maintenance of minimum levels must be taken. In addition, the EU Member States must ensure that the proportion of permanent pasture did not decrease significantly, compared to the 2003 determined proportion. (Agrarrecht)

Modulation

In addition to the production ("first pillar") measures of rural development ("second pillar") will be supported financially stronger. So far, in Germany will be granted a voluntary modulation. Therefore, direct payments are reduced by 2% annually. The, from 2005 compulsory modulation rates are 3% in 2005, 4% in 2006 and each 5% in 2007 to 2012. Every farm has an allowance of 5,000 Euros. The modulations are available for the first time in 2006, and are used for the development plans of countries to reinforce the actions of the 2nd pillar. The decisions on the final use of funds are made by the federal states within their respective programs. (Agrarrecht)

Reform of the Common Agricultural Policy (CAP) after 2013

From 2014, a new long-term EU budget comes into force, which will contain also a reformed CAP. The main topics discussed in the forthcoming negotiations are:

1. The increased promotion of common property, such as biodiversity and water, from agricultural subsidies
2. The extension of the so-called co-financing to let participate the Member States in paying the costs of subsidies
3. The redistribution of subsidies among the member states and between farmers
4. The reduction of the agricultural budget

In November 2009, leading agricultural economists from all over Europe gave a statement in which they demanded a clear focus on Europe's common goods, especially for climate protection, biodiversity and water. (Zahrnt, Reform the CAP - Hervest a better Europe, 2009)

4. The EU sugar policy

With a yearly output of 17 million tons, the EU is by far the largest sugar beet producer in the world. Therefore, it is no surprise, that sugar produced from sugar beets is one of the most subsidized crops by the CAP. Europe therefore is competing to levels produced by Brazil and India, the two largest producers of sugar from sugar cane. (American sugar alliance)

Because reforms concerning the raw material sugar were neither integrated in the 1992 MacSherry reform, nor in the 1999 Agenda 2000 decisions, sugar was also subject to a phase in 2009 under the "Everything But Arms", an initiative of the European Union under which all imports, of course with the exception of armaments, to the EU from the least developed countries were duty free and quota free. In 2005, the European Union agriculture ministers announced tactics to decrease the minimum beet price by 39% from 2006, over four years. Under the Sugar Protocol, to the Lome Convention, a trade and aid agreement between the European Community and 71 African, Caribbean, and Pacific (ACP) countries, nineteen ACP countries export sugar to the EU, and will be affected by price reductions on the EU market. (European Commission - Agriculture and Rural development); (About ACP sugar, 2005.)

These mentioned proposals of the World Trade Organization appellate body, mostly upholding on 28 April 2005, were the first determination against the EU sugar regime. (Agritrade - ethnical Centre for Agricultural and Rural Cooperation ACP-EU)

On the 21 February 2006, the European Union has made several reforms concerning sugar subsidies. The bonded price of sugar is to be decreased by 36%, with European output proposed to go down acutely as an effect of this. According to the EU, that is the primary real reform of sugar subordinated of the CAP since 40 years. (Europa - Press released RAPID, 2006)

One intention of this insurance policy change is to permit easier and more profit-making access to European markets for developing economies. Critics, say that this is neither a selfless move nor an idealistic shift from the EU, who are instead acting only in accordance with the wishes of the WTO, who supported challenges on sugar dumping by the EU from Australia, Thailand and Brazil. A further argument is that those countries that currently get privileged treatment from EU member states, often due to colonial ties though, as part of the ACP group may stand to lose out. (Ayliffe, 2006)

The EU Common Market Organization of sugar constituted in intervention leverage system in order to constitute minimum support prices and sugar secure. That sugar produced for domestic needs, carrying a production levy of 2% is called A sugar, the sugar produced for export with subsidy though, carrying an output levy of 37.5%, is called B sugar. The production prices are higher up than them on the global markets, but the the price support is limited to these two categories. Just quota suger should be used in the European Community. Sugar, which is produced in abundance of A and B quotas, is called C sugar and is exported without any direct export subsidies. It needs to be sold globally without the stand of export subsidies although the total European needs for sugar are even less than the sugar quotas A plus B together. The exports consequentially overproductions with export repayment and distorts therefore the worldwide trade flow.

The huge network of prejudiced tariffs and generalized country-specific or even region-specific business preferences certainly reflect the huge organization of patronage agreements which the European Union has constituted since a long time. Therefore, the EU sugar trade policy applies distinctive policies to various regions and a change of their direction from the principles of the World trade organizations is not dismissible. The EU has though a complex hierarchy of trade arrangements with specific groups of countries via the EU sugar regime, including the WTO, EU enlargement, African, Caribbean and Pacific (ACP) Countries, Least Developed Countries [LDCs], Overseas Countries and Territories (OCTs) and the Western Balkans.

The rules of the World Trade Organization (WTO) and the General Agreement on Tariffs and Trade (GATT) submitted the agricultural trade the first time in January 1995. The WTO agreement therefore mainly referred to commitments of three divisions namely market access, domestic support and export competition. Thus, between 1995 and 2000, the developed countries were obliged to implement the following commitments already, whereas the developing countries had even time until 2004. The reduce the import tariffs by 20%, limit the volume of subsidized exports by 21% and limit the amount of money spent on these export subsidies by 36%.

The reduction in the volume of exports and expenditures for export subsidies did not cause any difficulties for the yet. Contrariwise, the EU is even able to stay within the export subsidy commitments if they cut their production quotas yearly. That would be absolutely necessary, if the EU is in danger of infringing the commitments for export subsidies.

In the framework of the further liberalization agreement, the EU agreed to discontinue its export refunds on the condition that their trading partners also desist from their export subsidy programs. Concerning the matter of market admission, the EU is obligated to open the market.

The accession of the countries due to the EU enlargement probably results in a greater contribution to EU sugar consumption rather than production. Before these countries joined the European Union, they satisfied their sugar consumption via imports from Australia, Brazil, Cuba, Guatemala, Mexico and Nicaragua. (van Berkum, Roza, & van Tongeren, 2009)

5. Conclusion

Actually, first of all, there is something I would like to mention: I choose this topic because I found out that the enterprise *Südzucker* is Europe's number one receiver of agricultural subsidies and I thought, therefore, it would be rather easy to give them many hints to keep this position or even get more money. Then however, in the further observation of this topic, I came to the conclusion, that in my opinion a stock market dealing enterprise like *Südzucker* should not be one of the biggest subsidy receiver (or rather the common stockholders of *Südzucker*, what makes it even more disputable) There are loads of small and struggling farmers in the European Union who are lasting in terms of the environment, work species-appropriately with their cattle and also work highly economically but don't have a good position because of milk quotas etc. Rather they, for instance, should receive the lions share instead of huge stock corporations who are actually financially highly sound (*Südzucker* has been run according to their own profit and loss account in the 2008/2009 financial year a surplus of 183.2 million euro but gets a check from Brussels for well over 34 million Euros - even if excluding these 34 million Euros, would remain a surplus of 148 million Euros. The shareholders would then indeed might not get 86 cents per share more, but perhaps only 70 cents) From my point of view there is not really an urgent requirement that really maintains that huge amount (even the biggest, like I already mentioned) of money from the European Union. In the end, the tax-payer finance the dividend yields of the stakeholders.

In my opinion, subsidies must be given to companies that have been disadvantaged for any reason or no fault of their problems. Or, when a lot of jobs are dependent on that company or branch. When sugar from abroad is cheaper, then it can be quite profitable for the EU, to subsidize these companies, instead of the affect, that workers may need to get paid unemployment benefits for years to come. That in fact, would just be the principle of solidarity. Because of the hardly sources to get to know exacty why Südzucker gets these amounts, this point though is not applicable for me. Unfortunately the task for this assignment was to point out the factors that could be crucial for *Südzucker*. Hopefully Wolfgang Heer considers to pass a part of the subsidies to the raw material producing farmers.

As mentioned before, to increase its turnover of currently 5,9 billion Euros, the enterprise do not necessarily increase prices or even the amount of production arbitrarily like a furniture-producer.

Although the company got a lot of money in the last time, there is a trend of the European Union to decrease the agriculture subsidies in order to decline therefore costs of the first pillar of the Modulation policy. A couple of years ago, there was a yearly set of minimum price and if the market price was too low, the EU stepped into the breach. Because this however supported overproduction, it was replaced by income support. It certainly needs to be recommended to invest just in productions within the EU borders to get subsidies from the EU itself. Also advisable would be to stay nearby the adopted threshold price to gain the maximum of money. If possible, *Südzucker* has a chance to release an ecological product range certainly with ecological produced raw material because the EU rose the financially-aid-amount for that niche- but upcoming market. If it will be beneficial anytime in the future to import sugar from sugar canes of developing countries, it would be a chance, to save money because of small prices. The image not to produce sugar by supporting the rural agriculture thus, would probably be damaged though. Overproductions should be avoided in order to produce only A and B sugar and to get subsidies for it and there will be no necessity of exporting without getting subsidies for the products. The EU enlargement is a good chance for the company because the consumption is going to be bigger than the production. This could also lead to a higher production opportunity and thus to more profit and more subsidies.

Closing, it is recommended not to trust on subsidies but rather to concentrate on the business-development to be independent from any policy made in Brussels.

Bibliography

European Commission - Agriculture and Rural Development . (2003, June 26). Retrieved January 5, 2010, from CAP reform - a long-term perspective for sustainable agriculture: http://ec.europa.eu/agriculture/capreform/index_en.htm

About ACP sugar. (2005., January 28). Retrieved January 7, 2010, from http://www.acpsugar.org/old/

Agrarrecht. (n.d.). Retrieved January 6, 2010, from Agrarreform konkret – so werden die Reformbeschlüsse in Deutschland umgesetzt: http://www.agrarrecht.de/download/umsetzungAR.pdf

Agritrade - echnical Centre for Agricultural and Rural Cooperation ACP-EU . (n.d.). Retrieved January 7, 2010, from http://agritrade.cta.int/en/#sugar

Amercan sugar alliance. (n.d.). Retrieved January 7, 2010, from Foreign Supplier Profile: European Union - "Reform" of the Sugar Regime: http://www.sugaralliance.org/the-sugar-beat/foreign-supplier-profile-european-union.html

Auswärtiges Amt Deutschland (Federal foreign office). (2009, September 4). Retrieved January 5, 2010, from Die Gemeinsame Agrarpolitik (GAP) (The common agricultural policy) : http://www.auswaertiges-amt.de/diplo/de/Europa/Aufgaben/Landwirtschaft.html

Europa - A constitution for Europe. (n.d.). Retrieved January 2010, from The Institutions of the Union/The Minister for Foreign Affairs: http://europa.eu/scadplus/constitution/minister_en.htm

Europa - Press released RAPID. (2006 , February 20). Retrieved January 7, 2010, from CAP Reform: EU agriculture ministers adopt groundbreaking sugar reform : http://europa.eu/rapid/pressReleasesAction.do?reference=IP/06/194&format=HTML&aged=0&language=EN&guiLanguage=en

European Commission - Agriculture and Rural Development. (2003, June 26). Retrieved January 6, 2010, from CAP reform - a long-term perspective for sustainable agriculture: http://ec.europa.eu/agriculture/capreform/index_en.htm

European Commission - Agriculture and Rural development . (n.d.). Retrieved January 8, 2010, from Reform of the sugar sector: http://ec.europa.eu/agriculture/capreform/sugar/index_en.htm

European Commission. (2009, December 17). Retrieved January 5, 2010, from Financial Programming and Budget: http://ec.europa.eu/budget/budget_detail/current_year_en.htm

European Parliament. (2000, June 28). Retrieved January 6, 2010, from The Treaty of Rome and Green Europe: http://www.europarl.europa.eu/factsheets/4_1_1_en.htm

FAZ. (2009, January 14). Retrieved January 5, 2010, from http://www.faz.net/s/RubF3F7C1F630AE4F8D8326AC2A80BDBBDE/Doc~E08F0FAE281B1459A95E84 2386DD48267~ATpl~Ecommon~Scontent.html?rss_googlefeed

Finanznachrichten. (2009 , June 25). Retrieved January 5, 2010, from http://www.finanznachrichten.de/nachrichten-2009-06/14262022-dgap-adhoc-suedzucker-ag-mannheim-ochsenfurt-deutsch-016.htm

15

Frankfurter Allgemeine Zeitung (FAZ). (21. July 2008). *EU bietet niedrigere Agrarzölle an*. Abgerufen am 28. December 2009 von FAZ.Net: www.faz.net/s/Rub0E9EEF84AC1E4A389A8DC6C23161FE44/Doc~EBA4EB7F3A17E436BABD54FBB76 C9DAD6~ATpl~Ecommon~Scontent.html

German government for food, agriculture and customersecurity . (2008). *Miteinander von Nahrungsmittel- und Energiepflanzenproduktion ermöglichen*. Abgerufen am 22. December 2009 von www.bmelv.de/nn_750584/DE/04-Landwirtschaft/Agrarmaerkte/Rohstoffkongress.html_nnn=true

German Government of Finance (Bundesministerium der Finanzen). (2008). *21th subsidy-report of the German government about developement of financial aids - 2005-2008 (Einundzwanzigster Subventionsbericht der Bundesregierung über die Entwicklung der Finanzhilfen des Bundes und der Steuervergünstigungen für die Jahre 2005-2008*. German Government of Finance (Bundesministerium der Finanzen).

Handelsblatt. (2008). Retrieved January 5, 2010, from http://script.vhb.de/tabellen/html/egfl_subventionszahlungen.php?show=25&site=1&nr=6&sort=2

hri.org. (n.d.). Retrieved January 6, 2010, from http://www.hri.org/docs/Rome57/Rome57.txt

Südzucker . (2010). *Südzucker corparate homepage*. Retrieved January 5, 2010, from http://www.suedzucker.de

Südzucker. (2010). *Südzucker corporate homepage*. Retrieved January 5, 2010, from http://www.suedzucker.de/de/Unternehmen/Vorstand/

Tagesschau. (2008, Mai 20). *Das Ende der Subventionswirtschaft (The End of the economy of subsidies)*. Retrieved December 28, 2009, from www.tagesschau.de/wirtschaft/agrarreform4.html

van Berkum, S., Roza, P., & van Tongeren, F. (2009, Mai 6). *AgEcon Search - Research inAgricultural & applied Economics*. Retrieved January 15, 2010, from Impacts of the EU sugar policy reforms on developing countries: http://ageconsearch.umn.edu/bitstream/29139/1/re050609.pdf

Zahrnt, V. (2009). *Reform the CAP - Hervest a better Europe*. Retrieved January 7, 2010, from A Common Agricultural Policy for European Public Goods: http://www.reformthecap.eu/posts/declaration-on-cap-reform

Zahrnt, V. (2009). *Reform The GAP - Harvest a better Europe*. Retrieved January 7, 2010, from http://www.reformthecap.eu/